The Metal Moon

by

Everett C. Smith And Roman F. Starzl

Double9 BOOKS

The Metal Moon

by Everett C. Smith And Roman F. Starzl

ISBN: 978-93-59325-77-4

Published by

DOUBLE 9 BOOKS

2/13-B, Ansari Road
Daryaganj, New Delhi – 110002
info@double9books.com
www.double9books.com
Tel. 011-40042856

ABOUT THE AUTHOR

Roman Frederick Starzl (1899-1976) was a writer from the United States. He and his father (John V. Starzl) previously owned the Le Mars Globe-Post newspaper in Le Mars, Iowa. Roman Frederick's father was physician Thomas E. Starzl. His work is largely forgotten today, yet he was dubbed a "master" by E. E. Smith, the pioneer of space opera. Smith's Galactic Patrol may have been influenced by Starzl's Interplanetary Flying Patrol in The Hornets of Space. In Eric Leif Davin's Pioneers of Wonder, Thomas Starzl talks extensively about his father. He was the son of John (born Johann) V. Starzl and Margaret Theisen, and was born Roman Frederick Starzl. The son of Josef Starzl and Magdalena Ruba, John V. Starzl was born on April 9, 1865, in Bischofteinitz (after known as Vvrov) in the Austrian Empire. Josef Starzl moved to the United States in 1878 with his wife and five children (including John). Everett C. Smith is a diverse and skilled author noted for his ability to build interesting narratives across multiple genres. Smith's work captivates readers and takes them on thought-provoking adventures with a good sense of storytelling and an enthusiasm for exploring varied issues. He consistently produces engaging stories, well-developed characters, and beautifully conceived universes whether he delves into science fiction, mystery, fantasy, or other literary genres.

CONTENTS

The three men in the tiny space ship showed their apprehension as they watched the gravity meters. Something was distinctly wrong with the ship.

"Are you sure that there isn't some undiscovered moon of Jupiter?" asked the youngest of them. He was only about 25, which was very young indeed when his scientific attainments were considered, even for the human race's stage of intellectual development in 1,000,144 A. D. His figure was stocky, powerful, his face rather thin, bold, with piercing black eyes. He was naked, save for short, brilliantly red trunks of metalsilk. His name, "Sine," followed by a numerical identification code, was tattooed indelibly in thin, sharp characters on his broad, bronze-hard chest.

The man at the ampliscope removed his head from the eyepiece and shook his head impatiently. His body was bronzed and spare, but the complete absence of hair on his head made him look older than the 48 years indicated by the code following the name on his chest, "Kass."

"I tell you, Sine, this pull is no gravity effect. No body of such mass could be invisible, unless it were composed entirely of protons. And even then it would yank Jupiter out of shape, making it look like a pear, but there—"

Jupiter presented its usual appearance. The solar system's largest planet seemed enormous at this distance of only a few million miles. It showed its usual marked depression at the poles, but no distortion such as might be caused by a nearby body of enormous mass.

"What do you think, Lents?" Kass turned to the third occupant of the little space ship. Lents raised his broad placid face from the pad upon which he had been figuring a complicated equation. He was a large man, slow-moving, and fat. He was sensitive to that fact, so that, besides the usual trunks, he also wore a toga-like garment. His brown eyes blinked in folds of flesh.

"No doubt you're right, Kass," Lents rumbled in a deep voice. "I can't see how such a body could exist without pulling all of Jupiter's moons to itself. No, we seem to be specially honored by its attention."

They looked at one another soberly.

"The question is, can it out-pull us?" Sine remarked.

"You ought to know," Kass said. "You designed and built her."

Sine made his way forward. It was no longer necessary to use the handholds, for the pull of the mysterious body was already so powerful that it entirely eliminated the free floating so familiar to space travelers. Sine looked through the grated outlook windows, past the gracefully curved bow of the ship. At the very tip was the ether screw of his invention, resembling the screws used for water propulsion in ancient times, except that the pitch was extremely sharp. The tachometer showed that the screw had slowed down to 50,000 revolutions a minute, although the thermometer indicated that the molecular bearings were still reasonably cool. But how long could she stand the strain? How long, indeed, could the sturdy little atomic motor keep those blades turning? It was designed to pull directly away at a distance of only a million miles from the sun, and yet it was being beaten far out here in space by an object as yet invisible.

"What a crash that'll be!" Sine murmured, watching the agony of tortured metal.

Amidship, Kass was again studying the eyepiece of the ampliscope. Suddenly he stiffened.

"I see it! Why, it can't be over a couple of hundred feet in diameter. Cylindrical, I think. Head on to us now."

They crowded around him. Lents, with hasty computations, determined that they were still about three thousand miles from the object.

"No chance to pull away from it, if we pull straight," and his heavy voice was full of energy as his sleepiness vanished with the need for action. "Set her over, Sine, about 40 degrees. Try for a circular orbit around it—if we can get up enough speed centrifugal force will save us!"

Sine did as he was told, and the ship heeled over so that it presented its side to the sinister object, which was still invisible to the unassisted eye. While Kass watched it through the ampliscope, his companions stared through the thick ports at the velvet, gem-studded firmament. They could feel the attraction growing with terrifying speed.

"It's turning with us," Kass announced, "and getting closer. If we can swing around it, it will be a very sharp ellipse indeed!"

"Try and see if you can get a few more revs out of the screw," Lents suggested, and Sine crept forward, his powerful muscles straining against the pull. He lifted the leaden weight of his arm to the lever. He *must* get a little more power out of the motor, or they would crash to their deaths in a few minutes! A fine ending for their daring dash to Jupiter—the first space flight since the great comet swarm of 800,768 A. D.

Sine pulled back hard on the lever, and the motor gamely responded, moaned and shuddered under the tremendous overload. The tachometer needle quivered, began to climb, 52,000, 55,000, 56,000— —

The ship gave a lurch—there was a dull grinding, a hollow, metallic groan. The men picked themselves up from the floor—realizing at once the fatal significance of the lack of effort required. Their movement carried them off the floor—made them grasp handholds. Floating free! That meant falling free!

Sine glanced at the tachometer. The dead needle stood at zero. Through the forward window he could see one of the four screw blades, black, motionless.

Lents, obeying the habits of a lifetime, elbow hooked in a handhold, was figuring the time required for them to strike. He looked up with a puzzled frown.

"We should have struck about right now! Check on that body's position, will you, Kass?"

The bald-headed scientist pulled himself to the ampliscope. But it was possible to see the object through the ports now, quite plainly. It was black, cylindrical, glinting dully in the sun's light. The space ship was tumbling end over end, lazily, bringing the thing into view first at one port—then another.

"No acceleration!" Kass reported, amazement mingling with hope. "Same speed—we may still hit—but no evidence of gravity. We're falling toward it on momentum alone!"

Lents' brown eyes twinkled with perplexity in their pits of fat.

"The force, whatever it is, doesn't seem like anything in nature. But if we're traveling on momentum alone we can pull away with our emergency rockets—though I hate to waste the fuel."

Sine leaped to the rocket controls. "Grab handholds!" he snapped over his shoulder. The men rolled into the padded niches provided for that purpose. Sine's niche was so placed that it would not be necessary to lift a hand against the tremendous pressure of rocket acceleration. A lateral swing of the lever along its quadrant operated the rockets.

"Oof!" came a smothered exclamation from Lents as the ship seemed to pause, to leap forward in space again. The star-studded heavens as seen through the ports were hidden by a curtain of flame, electric blue and as stiff seeming as a steel bar—the trail of the forward rockets.

For some minutes there was no sound save the subdued thunder of the hull as it trembled under the tug of the rockets. Then a light flashed redly and a gong sounded. The signal that meant, "fuel half gone." Sine shut off the power, crawled out stiffly. His first glance out of a port showed that they were still falling toward the mysterious cylindrical space wanderer.

Kass wiped the sweat from his bald head.

"No use wasting any more effort," he said hoarsely. "That thing is a space ship, and there are men in it. The force they have been using on us is some kind of gravity beam—probably it's also their means of space propulsion. They mean to capture us, no doubt——"

"And they've reversed the beam!" Lents puffed as he turned away from the ampliscope, pulling his sweat-soaked toga away from his fat body with thumb and forefinger. "We're decelerating fast, but we can't feel it because the force acts on every particle of our bodies exactly the same as on the ship——"

"Proving," added Sine, looking out of the port curiously, "that it's a true gravity beam!"

The utter stillness of their ship gave the illusion that she was motionless, and that the sinister stranger was drifting toward them.

"It *is* a ship!" Lents rumbled. "Look at her ports. But they're shuttered."

"Not a bad idea," Sine agreed. "Protection against pin-point meteorites, anyway." They saw now that the cylinder was slightly rounded at each end, and the end presented to them had at its nose a circular projection, not unlike a very large button, that glowed with a lavender light, which they guessed to be the source of the gravity beam.

They were torn between the excitement of discovery and a very natural apprehension. In the dim past, more than 200,000 years ago, there had been a regular commerce between Earth and the Jovian colonies. But the comet swarm, coming out of the mysterious depths of space, had released to the solar system such swarms of meteorites as to make interplanetary travel in the spatial belt between Mars and Jupiter utterly suicidal. It required the passing of two thousand centuries to thin them out sufficiently to permit the voyage of exploration in which these three men were engaged.

What would these children of Earth look like after 200,000 years of Jovian evolution? Would they be friendly?

They must, at any rate, be curious people. The great cylinder was passing over them, and they had a better conception of its size. It was at least twice as big as the 200-foot diameter Kass had estimated, and fully 1500 feet long.

A section of its hull slid open, and the scientists felt the tug of mysterious forces on their own little vessel. They drifted up into the opening, knew that the hatch had closed by the shutting out of the solar glare. But there was no lack of light. They could see the welded plates of the hull by an intense saffron light that came from oval plates set in the wall. More of the gravity buttons were ranged around the room. It appeared that they were regularly used in handling freight. Now, as the little captive ship was tugged here and there, the prisoners could see flashes of that penetrating lavender light that seemed somehow solid.

"Get ready, men!" Sine said, breaking off his absorbed contemplation of their surroundings. "Strap on your belts, and be sure your disintegrator tubes are in their clips."

Lents was already lifting his toga and snapping his weapon belt around his ample waist. A mere strip of flexible metal with pockets for the atobombs and a clip for the delicate little tube—it might easily be taken for a mere ornamental article of apparel.

"Hope they're friendly," Kass remarked, patting the buckle shut over his flat diaphragm, "but if they aren't we can give 'em a thing or two to think about."

The quartz ports, kept free from frost on the inside by a curtain of hot dry air blown over them through a slit, suddenly misted over on the outside, became opaque with a milky glaze of frost. This told the prisoners that their captors were "bleeding" air into the hold, which did double duty as an airlock. They heard vague clanging of metal on metal, transmitted to them through the hull of their ship. Then a sharp blade scraped away the ice from one of the ports, and a face peered in.

They looked at one another for a few moments, these cousins of the human race, separated by 200,000 years of time and impassable meteor-strewn wastes of space. The man at the port turned and beckoned to others, who also surveyed the prisoners.

Then the first one, evidently the chief of this massive space vessel, motioned to the prisoners, to open their manports.

"Keep together now!" Sine admonished his companions. "If they act unfriendly we'll let them have the ray. Then you two slip back into your own ship while I grab this vacuum suit out of the lock. With that on I can carve a way out, and disable them, too."

"It would be a shame!" Kass said as he whirled the handwheel of the inner manport, "but——"

The valve opened, and a few minutes later the three Earthmen stepped out to confront the Jovians.

There were half a dozen of them, standing firmly, by virtue of the artificial gravity, somehow produced. They were not far different from Earthmen, except that they were shorter, being barely five feet tall. Their tremendous muscles told of the race's adaptation to the superior gravity of Jupiter. Their feet, encased in slippers of some burnished material, were unusually large.

They were dressed in an armor of overlapping scales that covered every part of their bodies, even their fingers. But their heads, instead of being armored, were protected by a thin, transparent membrane that followed the shape of their features closely. The Earthmen recognized the protective covering used before the comet swarm as a defense against the then used heat ray. So the Jovians had developed no new weapon! Sine thought comfortably of his little disintegrator tube. He could make those armored men vanish like puffs of smoke.

But they made no hostile move, and Sine had leisure to notice their faces. If their bodies were too heavily muscled for grace, their heads atoned for that defect. These were truly Jovian, god-like, combining intense virility, dominance, courage. But there was also about them an expression of intolerance, of ruthlessness, of selfishness. Here were men, it could be seen, who would not be too scrupulous in attaining their ends. But men, too, who could be charming companions.

Their leader, the man who had first looked into the port, now detached himself from the group and came forward, his hand outstretched in the old Earth gesture of friendliness. His appearance had all the characteristics of his companions, but in a more striking degree. He was taller than they, more than five feet, and his broad shoulders had the confident bearing of accustomed command. He spoke, in a pleasant, vibrant baritone:

"Welcome, men of Earth. Sorry for our little misunderstanding."

Sine gripped his hand, returned the muscular grip.

"It took us a little while to know what you were. And I may add that I'm pleasantly surprised that we can still understand each other."

The Jovian shrugged his shoulder:

"Canned speech. No chance for a language to evolve when it's mechanically recorded. But come up to my cabin. It's chilly here, and your manner of dress— —"

"*That* has changed!" Sine smiled. "Lents and Kass, will you go ahead?"

CHAPTER II
THE PLEASURE BUBBLE

After the first suspicions had worn off, the Earthmen felt that they had been singularly fortunate. To be captured by these intelligent beings had been about the most convenient thing that could happen to them. They might have found the human race entirely wiped out on the gloomy planet. Or they might have been struck by one of the still inconveniently numerous meteorites which would mean, at the very least, being marooned. Had they possessed the ability to look into the future they would not have rested quite so complacently in the hammocks assigned to them in the great patrol ship.

The big Jovian, they learned, was chief of the ship. He told them his name was Musters, and introduced his officers. They were an intelligent, efficient lot. From them the Earthmen learned something of the social organization of the human race as it survived on Jupiter.

"The race followed its natural evolution," intelligent and handsome young Lieutenant Reko explained to Sine as they leaned against a railing and gazed out of an unshuttered port at the somber splendors of Jupiter as it gradually swelled and covered the firmament.

"Like mated to like, and so the superior individuals became more superior, and the inferior ones more inferior. This resulted eventually in two races. Naturally we took steps to properly segregate the inferior race. Our efficiency experts have found ways to put them to work—to make them quite useful in fact. Of course we could not trust them with our weapons, our ships, our really important central power plants——"

What were these inferior—these so-called *Mugs*—what were they like? Reko arched aristocratic eyebrows. Why, they were often quite human in their appearance—though occupational diseases, and so forth——. Sine gained the impression that they were kept out of the way in order not to disturb the esthetic comfort of the superior race.

"There was a time when we had trouble with them," Lieutenant Reko said. "There were trouble makers among them. They attacked the homes of the First race, seized power control stations. Not fifty years ago there was an insurrection. But the Mugs lost. Thousands upon thousands of them were driven into the swamps and caves on the edge of the Tenebrian Sea. They were never seen again, although we searched for them with our heat rays. Perished, no doubt."

None were left now, Reko said, except those actually and fully occupied at certain labors for which they were found efficient. They were allowed to reproduce in sufficient numbers to fill the requirements—no more.

"What a rotten fate!" Sine exclaimed.

"They are quite a terrible people," Reko pointed out, closing a distasteful subject.

A few sleep periods later Musters called his terrestrial guests to his cabin.

"I have a pleasant surprise for you," he told them in his musical baritone. "Our planetary conference would wish for me to give you a most pleasant impression of Jupiter, so that interplanetary relations may be resumed under the best possible conditions. For that reason I am going to land you on a satellite that I'll wager will be a revelation to you. It is the goal and object of every one of our people. But it is costly and only a small portion of our population can be accommodated at a time. You may judge the kind of place it is by the name the public has given it: 'The Pleasure Bubble.' Come to the astrogator's cabin now; I'll show it to you."

They followed Musters to a compartment in the rounded bow of the great ship, stared out of a quartz port between opened shutters.

They saw Jupiter, immense, formidable, a mass of turbulent vapors, a depressingly drab scene. Suddenly Lents exclaimed, incredulous;

"Look! A satellite! There is no satellite this close to Jupiter! It's mathematically impossible!"

Musters laughed jovially. "It's there, isn't it? That's Jupiter's tenth satellite—The Bubble. It is less than 100,000 miles from the vapor envelope and has to travel so fast that its period is less than 8 hours. It was built by the First Race and set on its orbit so that our people would have a place where they could enjoy the sun, which is never seen from Jupiter's surface."

"It *is* a bubble!" Kass remarked, after an absorbed study of the satellite. It was racing just beneath them, at a dizzy speed, like a bubble blown before the wind. The ship followed the satellite, drawing closer, so that it grew in size and beauty.

Lents was mentally calculating the rupturing pressure exerted by the atmospheric pressure inside the crystalline ball. He stopped aghast at the thought of the tremendous strain.

"That crystalline material stands the strain easily," Musters assured them. "It will resist anything but a direct hit by a very large meteorite. As you can see now, the sphere, which is about a mile in diameter, is bi-sected by a plane surface, on which the city is built. In that little area you will see reproduced the choicest conditions of Earth." He turned earnest, hungry eyes on them:

"You don't know how lucky you people of Earth are!"

The ship was now coming quite close to the vast curve of the crystal, and they could see glimpses of beautiful structures in fairylike colorings, of small lakes like exquisite gems, of brilliant bursts of light that they conjectured served as substitutes for the sun while it was occulted by the enormous bulk of the planet.

Steadily the ship swept downward, to the level of the city, and the Earthmen became aware that the entire sphere was not transparent crystal. The part below the city level was a dull, ugly black.

"That's where the machinery is," Musters answered their questions, somewhat shortly, it seemed. "Hydrogen integrators there—to generate the power. Leakage of injurious rays down there—couldn't expect the First race to work there."

"Who *does* run the machinery?" Sine asked curiously.

"The labor Mugs, of course!" And Musters changed the subject.

The chief left them to their own devices as he superintended the lining up of the big ship's airlocks with the lock gasket of The Bubble. This effected, he bid his guests courteous farewell, assuring them that their ship would be conveyed to the Jovian capital city of Rubio, where they would be given every facility for repairing their damaged motor.

Sine was awakened by the talking of Kass and Lents as they sat at their breakfast in their unimaginably luxurious apartment. They were near the top of one of the fairylike towers they had glimpsed, and through the crystalline roof they could see the blackness of star-studded space. Far above was the glint of slanting sunlight on the outer covering of the sphere. This was the fourth morning on The Bubble, and the Earthmen were beginning to become vaguely restless. Their hosts had entertained them royally, but—

"I didn't see anything funny about the way they shoved that labor Mug out of the airlock," Lents was saying. "The poor devil! Stole a little of the juice they call ambrosia. The way that elegant over-civilized crowd laughed!"

"They lined up and watched the body floating alongside," Kass added somberly. "And that Mug was as human as you or I."

Their words recalled the scene vividly to Sine's mind. The broad, green field between two crescent lakes. The beetling-browed wretch, with eyes full of fear that darted from side to side, led to the center of the field by two splendidly armed warriors, there to be left alone in an agony of uncertainty.

He saw again the half-hundred clean-limbed athletes, sons of rich Jovian families. They were lined upon each side of the field. At the signal they dashed in. The frightened labor Mug tried to escape. As one team closed in he doubled, ran directly toward the others, saw his mistake too late. There was a brief savage scrimmage, and the unfortunate victim was stretched unconscious on the sward, while the victors and the vanquished in this curious game joined arms and made for the baths where exquisite nymphs peered coquettishly from behind delicately proportioned columns. Sine reaped uncomprehending and resentful stares when he declined to join them.

"Too rich for my blood," Sine told his companions at breakfast as they discussed their experiences. "Hope they take us to Rubio soon. We've done our job, and as for me, I'm not cut out for high Society."

After they had completed their breakfast a girl came hesitatingly into their chamber. Sine stared at her curiously. She had none of the enameled beauty of the women he had seen until then, but in her young face was a subdued comeliness that was attractive after the assertive pulchritude that was universal among the young women of the First Race. Unlike the shrewd display of their chiseled perfection, this girl's slender, rounded body was wrapped in a thin, gray garment that concealed as it draped. It was caught by a cord around her waist. Her feet, smaller and more fragile than the sturdy Jovian standard, were encased in neutral buskins. She stood submissively, waiting for them to speak.

"What does that girl want?" Kass murmured aside. "My stars, she can't be a labor Mug!"

"Come here, girl!" Lents rumbled kindly. "What can we do for you?"

The girl came forward hesitatingly. Her voice was soft, lacking the brassy assurance of other Jovian women;

"I was sent here, masters, to guide you through hell."

Immediately after this startling statement her face turned a brilliant red, then a deathly white. She half turned as if to flee, but, as if realizing the uselessness of flight, she faced them again, defiantly;

"I don't care what happens to me!" she declared desperately. "I've told the truth at least once. Jovians call this place The Pleasure Bubble, but they don't have to live in the black half. Now tell them what I have said."

"We will not tell anyone what you said, child," Lents rumbled comfortingly. "But tell us. You don't look like the Mugs we've seen so far—nor like the poor fellow we saw put through the airlock. They seemed—a different race. But you—why—on Earth we could hardly tell you from any other kid of your age."

A flash of spirit illuminated the girl's tragic, immature face.

"They call us a different race!" she exclaimed. "True—but not an inferior race! They are the inferior race, though the stronger. They depend on our knowledge, our labor, to live! My father told me so!"

Kass, who had been studying her silently, asked, "Your father?"

"Yes. The technic in charge of the machinery below. He was ordered to escort you around. But his scars from the rays make it hard for him to breathe today. He is in his bunk. So he sent me in his place."

Sine wondered if life under such unnatural and destructive conditions would some day reduce this graceful girl to a horrible parody of humanity. He asked;

"Do you work below?"

Her clear gray eyes fell on him.

"No. I was selected by the Committee to work in the Baths when I am sixteen. I am fifteen now."

"Holy twisted nebulae!" Sine swore under his breath. "The kid doesn't know what her work in the Baths is going to be! So the Committee selected her for the Baths!" He felt suddenly a violent dislike for the very rich Jovians, a feeling of fraternity with the Mugs.

"We will be very glad to have you guide us," he said formally. "What is your name?"

"Proserpina. My father said it is fitting for one who lives where we do."

Strange anachronism! That name from the mythology of Earth's youth. Like that goddess of the underworld from misty antiquity, she led them down, down, until it seemed they must be near the bottom of the black hemisphere. It was a world of dim distances, of shadows, of pipes and girders, or grisly abysses from which came mysterious sounds; of locked chambers in which ghastly fires flared.

Now and then they met the inhabitants of the place; misshapen Robolds going about unknown tasks. They stumbled suddenly out of unnoticed passages, carrying burdens, grotesque, apelike, weary. Most of them were hideously deformed.

Several times, when their journey led them into a certain part of the hemisphere where they felt strange tingling of their nerves, the girl led them away.

"We must not go there," she told them. "The integrators are there. There my father received the scars of his chest that keep him from breathing. Most of those who are blind worked there."

The Earthmen had already heard hints of the atomic integrators from which the Jovians obtained endless power. They had no desire to get too near those searing by-products of power.

"Do you mean to say," Lents asked, puffing a little from their exertions, "that people down here live here all their lives?"

"I will show you our home," Proserpina said simply.

They came to it presently. A niche, a metal-laced nook, deep in the hull. Gigantic girders formed one side of it. On the other side enormous air conduits. It was clean, bare, not as depressing as they had expected. It was more like a gallery, long and narrow, sparsely furnished.

Something rolled out of a bunk at the farther end. Something like a great spider. A man, stooped over, his once powerful body doubled, so that his knuckles almost dragged on the floor-plates. He came toward them, fierce gray eyes looking out at them under bushy brows. So formidable that Sine's muscles tensed.

"Are these the visitors, Proserpina?" His voice was husky, as though his constricted chest with difficulty performed its function. He looked at them intensely.

"They tell me you are from Earth. Are you with us or against us?"

"Father, be careful!" She put her hand over his mouth, to be shaken off impatiently. But the girl's warning had taken effect. The man—it was impossible to tell if he were old or young—looked at them broodingly.

"My mother died here," Proserpina said. "And I am afraid he will. His mind is not as clear—"

Lents, distressed to the bottom of his generous soul, helped the victim of the Jovian pleasure moon back to his bunk. "This girl," he muttered to Kass, "can't we get her out of here?"

He had not meant for her to hear, but her quick ears caught his words, and a ray of hope illuminated her features. She was standing beside Sine, and her thin fingers gripped his hard bronzed arm;

"Oh, could you take me away? I will be your slave!"

Sine gently disengaged her fingers. He was strangely embarrassed.

"I'd like to. But I'm a bachelor man. No place for you, you know."

She did not persist. No doubt she realized that she could not leave that gaunt parody of a man who was her father.

When they bid farewell to Proserpina they were steeped in profound depression. Alone in their room, they talked over what they had seen, but they could think of no way to save Proserpina from her fate. They were still discussing their visit when the manager of this satellite of delights called on them and informed them that Governor Nikkia of Jupiter awaited them in the capital city, Rubio. A space ferry was even then clamped to the locks to take them to the mother planet.

CHAPTER III
THE COMING OF THE TEARDROPS

Governor Nikkia was like the majority of the First Race. Although he was not large of stature, his powerful muscles bulged impressively under his clothing. The two relatively slender Earthmen, naked save for their trunks, looked almost ridiculously puny. Lents' portly figure was more impressive, but the big scientist had all he could do to carry his weight, so uncomfortably augmented by Jupiter's great mass. The unaccustomed thickness of the atmosphere, too, made the Earthmen uncomfortable. The heat was excessive, for although the outer cloud masses had been determined by photometric telescopic examination to be near the freezing point of hydrogen, Jupiter's enormous store of internal heat made its surface temperature average around 100 degrees Fahrenheit. The humidity was high, and the explorers from Earth were distressed.

Nikkia was a good host, however. He ordered out one of the government cars, luxurious conveyances supported by gravity repulsion buttons, and personally accompanied his guests on a tour of inspection through the murky fog. They rode interminably over wet, domed roofs, down through gloomy arcades. Thunder rumbled incessantly, and occasionally there came a lurid glow of lightning.

For a city of Rubio's extent, they saw very few people. Occasionally they saw the erect, confident figure of a member of the First Race, tending some mighty engine whose purpose they could only guess. The inhabitants preferred to stay indoors, if they could not afford to dally in The Pleasure Bubble.

Nikkia listened with interest to the voyagers' account of their journey through space. But he did not respond with much enthusiasm to the suggestion that interplanetary commerce be resumed.

"We are comfortable," he said good-naturedly. "Besides, I'm not sure that the Mugs could build ships suitable for such long trips. They're getting lazier every day!" He shook his head regretfully.

"What do you expect?" Sine blurted. "You treat them like slaves, ruin their lives, and then you're surprised because they lack ambition!"

Nikkia looked at him in mild astonishment. "But they have to be kept in their place! If we gave them free hand they'd soon run us out. Why, not fifty years ago——"

He told again of that uprising that had resulted in the breaking of the Second Race's pretension. "We have to control 'em," he ended smugly.

The Earthmen were baffled by the bland indifference of the Jovians to their mother planet. They met many of the First Race in the next few days, but none seemed interested but the so-called Mugs, the Second Race, and their interest was wistful akin to nostalgia.

But the three scientists were to learn that the First Race were good fighting men, regardless of their short-comings in other lines.

The glowing "teardrops" appeared a little over a week later. They were so called because of their shape, but the Jovians knew as little about their nature as did their guests. They appeared early one murky morning, as Kass, Sine and Lents sat at breakfast with Governor Nikkia. The servants, comely, characterless specimens of the Second Race who held themselves snobbishly above their fellows, came panic-stricken;

"Your Supremacy!" called one, making a low obeisance. "There are strange lights hanging over the palace!"

Nikkia brushed the slight fellow aside, dashed up a stairway to a terrace on the roof, closely followed by his guests. In a few moments they were all soaked by the warm downpour as they stood on the terrace, like an island in a sea of brown fog.

There were three of them, roughly egg-shaped, but with an elongated tail. More like tadpoles, save that the tail was rigid and emitted a fiery streak. Obviously they were propelled by a new adaptation of the old rocket principle. They swam back and forth slowly, as if questing for something, leisurely selecting their victims. The strangest thing about them, however, was the light. A brilliant red, almost pink, like the glow of a neon tube, it penetrated the fog. Its pulsations even penetrated brain and body, so that the watchers became unpleasantly conscious of it.

Nikkia, watching tensely, turned suddenly on his guests;

"Damned funny! Barely you show up, and now this! I don't like it. Are they from the Earth?"

Lents swelled in slow and ponderous anger.

"Do you think, sir, that we are of the sort to abuse your hospitality by spying on you? We don't know any more about those things than you do!"

"Damned funny!" Nikkia repeated to himself. "Wonder if there's any of *them* left?"

"Your Supremacy!" a servant interrupted. "Call from the war office!" He was carrying a drum-like contrivance, carried on a stand, and set it down in front of the governor.

"Well?" Nikkia snapped impatiently.

The screen which formed the drumhead glowed into life. A Jovian officer, looking exceedingly efficient and warlike in his armor uniform, stood at salute, which Nikkia returned impatiently.

"Who are those flyers, Sonta?" the governor snapped.

"I don't know, Your Supremacy," the officer growled. "They fail to answer our challenge, and none of the men have seen anything like them."

"Then why don't you turn the heat on them?"

"We have. Our heat-rays have no effect on them. That pinkish light is a reflector wave of some sort. Several of our beam projectors were burnt up by the kick-back."

"Ram 'em then! Ram 'em! Sacred Ganymede! Is our Defense Service degenerating into a crew of Mugs?"

The officer's image on the screen was seen to flush, to draw itself up resentfully.

"We have sent ships up to ram them, Your Supremacy. Three of them have been destroyed."

"I was watching. I saw nothing."

"The visibility is worse than usual. They are half a mile high. Our own ships are invisible at a hundred yards. It's that cursed light."

Nikkia shut him off peremptorily.

"Never mind the conversation, Sonta. Get out every available defense craft. Box those teardrops. Ram them. Destroy them—I don't care how!"

The screen was suddenly dark, and Nikkia gazed angrily up at the mysterious glowing craft overhead. So far they had done no damage except to the city's fighting ships.

"Listen!" Sine exclaimed. His body glistened like wet bronze as he stood in the half darkness and strained to catch some sound over the steady patter of rain. "Lents, quit puffing!"

From high overhead, some sounds were coming to them. A steady, droning rush, like the sustained exhaust of rockets. That must be from the visitors, for the official ships were equipped with the gravity buttons. Now and again one of the glowing teardrops would be thrown violently from its course, evidently the effect of impingement of the gravity beam. But not one was disabled. The defense ships were not faring so well. Every little while there would be a fog-muffled crash as one of them crashed, throwing a stone roof into the street. But none fell near the governor's palace.

It was uncanny. No sound save that low, sibilant roar, and an occasional crash out there somewhere in the darkness. The mysterious attacking ships so plainly visible and so immune, and the defensive fighting craft, flying in silence and invisibility—crashing anonymously.

Nikkia had dropped his air of assurance and calm superiority. He was frankly worried, and still a little suspicious of his guests. This attack—it did seem rather a coincidence. What would Sonta have to report now?

He twisted a dial on the side of the communication drum. A junior officer appeared on the screen.

"What the devil?" the governor exploded. "Where is Sonta? I'll have him broken for this! Lieutenant, call Colonel Sonta at once!"

"Your Supremacy," the lieutenant said respectfully, "Colonel Sonta went up in one of the guard ships, and it has been reported crashed south of the catalyst plants."

For a second Nikkia stared at the screen, then snapped the switch wordlessly.

The attackers seemed to have broken down the capital's defenses. Here and there, through the thick, greasy fog, a lurid red glow would take life. That was the fog-diffused reflection of a heat-beam, probing the sky for the "teardrops." After a little while the glow would flare up and as suddenly die down, followed by utter blackness. Another heat-beam out of commission.

Nikkia was frantically polling all of the city's defense commanders. They reported failure with monotonous regularity. The electronic barrage wall around the city had been passed easily—the equipment wrecked. A proton bombardment had yielded exactly nothing—He snapped the switch, peered eagerly at the mist curtain overhead—there was a series of heavy concussions. The glowing visitors were being bombarded from above. The screen glowed again....

"... but the bombs are all detonated long before they get in effective range of...."

Close by a vague shape—a darker shadow in the muggy air, suddenly materialized. It was falling swiftly—a familiar cylindrical shape with rounded ends—one of the Jovian guard ships. It struck scarcely a hundred yards from the palace—struck with a jarring burst of sound like rending metal. Then utter silence again, and darkness. No cry of wounded man. No man could survive that fall and live.

"Some kind of emanation—shields them from all known attack—" Nikkia swore monotonously and regularly.

The glowing ships now settled down to the real purpose of their attack. They began to course back and forth across the city, methodically. Like burning meteors they disappeared over the horizon, to the city's farthest suburbs, back again, as if over a measured and marked course.

And like burning, melting meteorites, they shed trails of sparks, blazing liquid. Wherever these fiery drops landed there ensued immediately a dry crackling, followed by the rattle of falling masonry. As none of the buildings were inflammable, there was no danger of fire. But wherever this incendiary trail fell, stone cracked and crumbled.

"They are destroying us! Forty million people live here in Rubio. They will kill us all, women and children too!"

"*Who* are *they*?" Sine asked suddenly.

Nikkia looked at him bleakly. "Who? Why, the Mugs, of course! Those we banished. Those we thought we wiped out."

"Oh, yeh." Sine's intonation was very dry. "They're giving you a dose of your own medicine."

Nikkia did not reply. As if he apprehended, too late, that his statement might have sounded like a plea for help, he shrugged his massive shoulders with elaborate indifference, saying;

"I and my wives are not afraid to die!"

The Earthmen could no longer watch this ruthless destruction, however, regardless of the provocation.

"You say that pink light is a protection against every *known* mode of attack?" Sine asked, turning sharply to the governor.

"Yes. And that's sufficient, isn't it?"

"Is it proof against *this*?" Sine jerked the little tube out of its clip, directed it against a stone parapet that loomed grotesquely through the fog. A brilliant white beam leaped forth, cutting the fog like a bar of platinum. Then there was darkness, and the governor, examining the parapet, noted with growing hope that a stone pillar, a foot in diameter, had been cut off smoothly, cleanly.

"The disintegrating ray!" he murmured. "I have read of that, in fiction. But here! Here it is!"

Suddenly he was all energy.

"Will you use this weapon against our enemies? I assure you that you will be well rewarded. As much eka-iodine as your ship will carry! My own ship is here, in the courtyard. It is swift, and powerful. You have already learned the controls. Take it. Bring down those murderers!"

The fiery meteor was coming toward them again, planting a swath of death a hundred yards wide. There was really only one answer possible. The terrestrial scientists, having come on a mission of peace and discovery, stepped forward in unison.

"Give me the activator key!" Sine said crisply. "Lents, will you see that the port gaskets are loose? Kass, I'd like to have you take the controls."

"Right! Right!" They ran past the governor of the greatest planet in the solar system, ignoring him, down the broad stairs, through halls of weighty magnificence, and into the rain-sluiced courtyard.

The governor's ship was waiting there. Not very large, but fine. Its polished metal gleamed richly.

"Quick, inside!" Sine threw open the manport valves. They were inside. The gravity buttons glowed with their peculiarly material lavender light, and the ship rose vertically with swift acceleration.

From the sky the death trails left by the invaders were clearly visible through the murk which obscured everything else—a pink, pulsating light. And the three glowing vessels were coming toward them.

"Get above them, Kass!" Sine commanded. "When they pass under I'll let them have it."

Closer and closer they came, those blobs of light. The Earthmen could see nothing but the light—get no hint of their construction. But that there were men inside they never doubted. The glowing ships seemed to swell, to expand monstrously, and their throbbing emanations became more furious. They seemed to hesitate as they were about to pass beneath.

“They see us?” Lents rumbled, pulling at his toga nervously. The cloth was soaked, clinging to his fat body.

“Close enough!” Sine decided, leaning out of a port, disintegrator ray tube in his hand.

At that instant the strange pink light seemed to encompass the whole planet. They were bathed in it. The fog was a sea of baleful pink. Sine stiffened into impotent rigidity. The ray tube fell from his numbed fingers. He felt himself floating, weightless, in a sea of red that smothered him deliciously. And swiftly even that consciousness was succeeded by black oblivion.

CHAPTER IV
THE MONSTROSITIES

"He's coming out of it. Hand me the water, Lents."

Sine awoke to see Kass bending over him. He felt weak and languid, and the memory of recent events was returning only slowly. He looked around, saw that he was lying in a chamber about fifteen feet square, evidently hewn out of solid rock.

"Are you all right, Sine? Answer me, boy!" Kass' bald head gleamed in the yellowish light of a single emanation tablet on the ceiling.

"I'm all right. Where are we?"

"Under the sea. Some hidden city of the Second Race—those that were banished. We are prisoners, but honored prisoners it seems."

Sine passed his hand over his eyes.

"How did we get here?"

"Some kind of emanation of theirs—the brightening of that light, I guess. It had a paralyzing effect. I know I froze where I stood, unable to move a step. And I was protected by the hull. Same with Lents. But you had your head out of the port—caught the full effect. It laid you out cold."

"They boarded us then," the fat man supplied.

"As easy as that! Simply boarded us, herded me and Lents into their own ship, which is just as suitable for navigating in water as in air. As for you, they had to carry you."

"Better tell him what to expect," Lents suggested.

Kass explained, with considerable scientific interest:

"The First Race was not so far wrong in calling them 'terrible people.' They are, a race of monstrosities. Men with four or six arms, men with hair like fur all over their bodies. With heads ten times too large. With boneless tentacles instead of limbs. With scales instead of skin. Quite horrible. And yet, most of them are highly intelligent, with normal human emotions, and painfully conscious of their deformities."

"I don't quite understand." Sine was flexing his muscles, sitting up with the support of one elbow. He saw he was lying on a pallet of dried sea weed. "What caused these abnormalities?"

"Well, you know—" Lents was speaking judiciously. "You know all about the mutations produced by X-rays in the biological laboratories?"

"Of course!" For approximately a million years these actions of X-rays had been understood—their ability to bring about extraordinary mutations in the life-germ, whether animal or vegetable—the acceleration of natural evolution a millionfold. "But you don't mean to say the First Race deliberately brought about these mutations in the Mugs?"

"Not deliberately. But they permitted it with utter callousness. You know those hydrogen integrators we saw at a distance in the dark half of The Bubble. Those things are the source of most of the power used by the Jovians. But the generators have a mighty dangerous by-product—the cosmic ray series, for instance, a particularly destructive band below the X-ray spectrum too."

Sine nodded comprehension, his eyes hardening as he thought of the grotesque, distorted wreck of humanity who was Proserpina's father. A mere whim of fortune that he had not been condemned to that hell before she was born, or she might have been one of those unfortunate mutations—

Might yet become one! Not only could the rays deform the offspring. They could distort the full-grown, normal body. Sine felt increasingly dismayed as he thought of this immature, quiet-eyed girl, this waif of an alien world. He experienced a recurrence of the indignation he had previously felt. This selfish, superior First Race! Condemning the weaker people to torture and death so they could enjoy a little paradise! The Pleasure Bubble they called it. Sphere of the Damned was better! For the unfortunate consigned to the dark hemisphere was condemned to an inferno that surpassed the Ancient's most perfervid imagination.

"I wish we could save Proserpina!"

The words were out before Sine knew it. Kass stopped in the middle of a sentence and lifted a quizzical eyebrow.

"Oh, get the romantic ideas out of your heads!" Sine snapped. "You know she's just a kid. I couldn't take care of her if we did take her back to Earth. But I'd like to take her out of The Bubble!"

Lents pulled at his toga thoughtfully. It was dirty, still wet, and smelled not too pleasantly.

"I could take care of her," he said slowly, and his deep bass voice was a little wistful. "My wife would be glad—we're getting old, and no children—"

"We-ell," Kass submitted practically. "I'd like to take her away, and her poor old daddy too—or is he old? But what's the use of discussing all that? Here we are prisoners, and she's a prisoner of the First Race, and we shall never see her again. Or the good old Earth either," he added sadly.

A man entered the room. He looked more like a normal man than might have been expected—only his exaggerated dish-face, his bulbous forehead proclaiming him just another victim of the First Race's industries. Or his shrill, treble voice as he announced:

"Gentlemen of Earth, the Manager and his council expect you in the office. Follow me." He turned, waited for them to come.

The Manager's messenger led them up a long, ascending tunnel meagerly lighted at intervals by small emanation tablets. After they had gone perhaps a hundred yards the hewn rock gave way to what was evidently a kind of concrete.

"This part of their city is built above the ocean floor," Kass remarked quietly. "They brought us in through airlocks. Passages lead to caves along the shore where the original refugees holed up. These are mostly their children, so marked and deformed even in embryo."

Their dish-faced guide now stepped aside as they entered a spacious chamber with a domed ceiling. Here and there it was wet. No doubt above there was the sea. Lents made a rapid mental calculation, rumbled into Sine's ear:

"Can't be so deep. Not over a hundred feet; maybe less. Otherwise those arches couldn't carry the weight."

A hush fell upon the room. The leader of this strange people—the one they called The Manager, was rising from his seat back of a desk. His head was very large, his eyes large, liquid and expressive. A total lack of eyebrows, of hair on his head, gave a mixture of the comical and the obscene to his appearance. But the respect with which his counselors, ranged on either side of him, regarded him, ignored his appearance. They were all, without exception, victims of the strange and terrible mutations of type induced by the First Race's callous disregard to the dangers of the rays. All wore loose garments of drab material which concealed their deformities to some extent.

The Manager's large, intense eyes fastened on the Earthmen, and he addressed them:

"Men of Earth: We have captured you in battle, but we would be friends with the Old World. Why did you try to fight us?"

"You were murdering helpless victims," Sine said shortly. "It was not our fight, but we could not stand by and permit such a thing."

Something like amusement flashed up in The Manager's enormous eyes, so old, weary and wise.

"So you could not bear to think of an easy death for those of the First Race? What think you of their treatment of us?" He raised a scrawny arm—so thin it suggested a skeleton. "Hunted like beasts—imprisoned and tortured! Are we not human?"

"You see," Kass interposed diplomatically—"we were their guests. And in a way their quarrel...."

The Manager cut him short peremptorily:

"You were their guests! You lolled with them in The Pleasure Bubble, in the beautiful sun! The sun that most of us have never seen! And down in the dark half-human beings like yourselves—toiled and slaved at those devilish integrators to keep the machinery of pleasure going.

"You were the guests in the Governor's palace—in the magnificent city of Rubio, though to you it may seem dismal. But did you think of the poor slaves, deep underground, in the slimy sewers, in the uranium pits, in the power plants? You basked in luxury with the First Race, and their fight was your fight—their enemies...."

He was working himself into a fury, evidently forgetting the original purpose of this conference with the prisoners. But one of the counselors now approached him, bowed respectfully so that his scaly face was hidden. The Manager cut short his tirade.

"What is it, Gnom?"

"Isn't The Manager digressing?" Gnom asked in a hollow voice. "These men of Earth are now our guests. They come at an opportune time—when we shall reap the fruits of our long planning. If we wrest power from the First Race, shall we not need the friendship of the Mother Planet? Let them, then, carry our story to Earth, if it be that we may need their help."

The Manager stood in thought. At last, coming to a decision, he asked sharply:

"With whom do you stand, men of Earth? With us or our oppressors?"

Kass and Lents looked at one another blankly. They started as Sine spoke up sonorously, beside them:

"Officially, we are supposed to be neutral. But if you attack The Bubble and rescue the poor devils in the dark hemisphere I'll help!"

But The Manager shook his enormous head slightly.

"That we can not do. That satellite is too far out in space. There is no concealment, and we can not yet fight their patrol ships in space."

"Listen!" Sine persisted. "There is a man there I know. He's about ready to die, unless he gets away. And he has a girl, a kid of fourteen or fifteen. The rays haven't made a freak out of her yet. I want to save her. Give me a ship and I'll take her out myself!"

"That we can not do. Individuals do not count. One, or a hundred, may die. We can not endanger our plan."

The counselors had drawn a little away from the Earthmen, unconsciously symbolizing their support to The Manager. Again he raised his bony arm.

"Up above there our ships are destroying every city of the First Race on the planet. Our power-beams for the glowing ships are encircling Jupiter in a network of red and death—death to the oppressors! The Pleasure Bubble's turn will come. And when it is dashed down, master and slave must die together. To save the slaves might let some of the masters escape."

"Gentlemen!" Kass was trying to smooth over the situation, "We have been sent here on a voyage of discovery, not of war. We regret your troubles here—but we can take no part in them. Our attitude is friendly to...."

"No! Damned if I will!" Sine shouldered his iron-hard body through the close-packed counselors, so that he stood directly before The Manager, who did not shrink from the formidable young man. "If you murder those poor Mugs in the black hemisphere, I'm your enemy from now on!"

"And I!" The words boomed and reverberated in the vaulted chamber, and Lents moved his bulky body beside Sine.

"And I too!" Kass' naked, skinny torso glistened with sweat. "The First Race may be murderers, but they're magnificent murderers. They wouldn't forget their friends!"

The Manager's large, liquid eyes seemed suddenly filmed over. He jerked his enormous head sharply, snapped:

"We waste time. Put these meddlers out through the locks, that they may feed the fish."

But Gnom again interposed.

"If The Manager will permit—there is much water on Earth. They may know how to swim—might go to the top and escape—"

"True, Gnom. I have a truly great brain, as all the oppressed admit, but details escape me. Call one of the watch, put them to death first."

Gnom turned, looked into one of the larger passages that centered on that room. He turned his blank, scaly face.

"The watch is not here!"

"Perhaps he was called. See!"

But before Gnom could execute the order a commotion arose in the passage. A voice called from outside:

"Officer of the hour prays audience with The Manager."

"Enter."

An officer with an extreme hunchback dashed in, bowed low before The Manager.

"It is the end!" he gasped. "They watched our glowing ships plunge under the water, and they are setting bombing rockets for this area. The first ranging shots have already been fired. Listen!"

After a few moments there came a dull thud, as though a blow had been struck against the ceiling. A pendent drop of water fell. The Manager's hairless face became bleak.

"I made great plans, great inventions—forgot a simple detail!" He slumped as he stood, a mixture of the absurd and the tragic. The mutation that had made a brilliant mind had nevertheless left it incomplete, and none had realized it until in this extremity. Again came that dull shock, and this time it seemed a little stronger.

The Manager shook off his apathy. His great eyes burned with livid fire, as he called:

"Officer of the watch. Take these prisoners to the locks. Kill them and put them out."

"I obey!" The officer, squat, with enormous torso, pointed a small wand, pointed with a tiny spot of that peculiar pulsating pink light, threateningly. Stolidly he herded them through a broad corridor. Now and then they passed inhabitants of this submarine city, nightmarish, pitiable creatures, now disturbed, dreading death. Sine wondered vaguely that they should cling to such an unhappy existence.

He was recalled to their own predicament when a metal gate, closed by a screw-wheel, loomed up in the poor light. The inside lock! The guard motioned them ahead, stood between them and the passage. He fumbled at his belt, ignoring the dull hammerblows of explosions transmitted by the water. He seized Kass by the throat, prepared to plunge the knife into his body.

Sine leaped past, crooked his arm around the man's thick neck, attempted to break his neck. But a giant arm threw him off easily. He fell to the floor. Like an echo came the concussion of another explosion.

The guard, without trace of ill-humor, turned his attention to Sine. He pointed the little wand at him, and the light glowed brighter. Sine felt again that torturing paralysis. His senses were leaving him. The pink light was throbbing, expanding....

He wondered why the stones of the passage should be pushing in, spurting water. The pink light faded. Tepid water struck him, stinging like needles. There was a roaring, blackness. A fat arm hooked around his waist—Lents', no doubt. He felt himself borne along in a swirl of water, strangling, fighting blindly. There was another terrific explosion shock, an interminable climbing struggle. Then his head broke water and he breathed air again. Lents came up beside him, puffing and blowing, and after a long wait—so long that they despaired, Kass came weakly to the surface.

CHAPTER V
THE STRUGGLE FOR FREEDOM

They were afloat, and comparatively safe from the rockets which shrieked out of the leaden sky and threw spectral waterspouts up into the fog before they exploded. Unless one exploded directly under them, or very near, they would be safe—for the time being.

"Which way is shore?" Lents puffed.

"Rockets seem to come from that way," Sine answered, flipping his hand. "Swim that way. Fish probably lost appetites, so won't bother us."

The bombardment had indeed frightened away the monsters of the deep, and even the dead in the ruined submarine city would rest in peace for a while. But the Earthmen, after several hours of swimming, doubted that this was more than a postponement of death. The long greasy swells were rising, presaging another of Jupiter's unimaginably violent storms.

"I see a light!" Sine strained his eyes to get another glimpse of it through the brown fog. "There it is again." Something was moving slowly through the air a short distance over the water, following the course of the rockets, which had ceased coming. A powerful searchlight was cutting through the murk. A war party of the First Race, looking for wreckage.

In their methodical search they soon found the swimming men, and they were helped into the chief's cabin. Sine, looking up with half-blinded eyes, saw Governor Nikkia sitting in his chair, looking at him coldly.

"So!" the governor bit off his words. "The traitors are fished out." His arrogant, handsome face was vindictive, uncompromising. "We forgot that the aborigines of Earth would naturally sympathize with their equals, the Mugs! That was nicely timed, your 'visit.' How long have you been in communication with the rebels?"

The Earthmen, weak and exhausted by their long exposure, resisted their desire to lie down on the floor. They stood before the governor, hemmed in by hostile fighting men, and tried to maintain the traditions and dignity of their planet.

"We were not in communication with your slaves," Sine declared. "You should know that. Your radio monitors would have picked up any messages, and your own patrol ships picked us up when we were far out in space. Our mission is one of peace. As for your quarrels, they do not concern us. We are strictly neutral."

Nikkia laughed, a short, clipped bark in which there was little amusement.

"Well, your guilt is a matter of small moment anyway. We have paid the Mugs for the damage they did, and they will not have another chance. And if they had an idea of getting help from Earth, you shall be an object lesson on the uselessness of such hopes."

"Meaning?" But Sine and his companions knew that the meaning must be evil.

"Meaning," Nikkia snapped, "that from now on you three are Mugs, no better and no worse than the Jovian Mugs. Except that I shall instruct the labor office to put you to work at one of the power integrators—perhaps in The Bubble. We don't want to waste you" he added with grim humor—"and the gravity here on Jupiter might reduce your life of usefulness."

The governor turned his back in dismissal, and the prisoners were hustled into a dark, extremely hot storage hold. Here they lay down amid an untidy collection of miscellaneous gear, thick with dust. They rested gratefully until some of their strength should return to them.

When they awoke from their sleep of exhaustion they were aware that the ship had landed, and a few minutes later the door of their prison was opened and an officer, heat pistol trained on them, commanded the prisoners to get into another ship for transfer to the metal and crystal satellite where they were condemned to drag out the rest of their lives as slaves.

The second coming of the Earthmen to The Bubble was in marked contrast to their first. Instead of the large, commodious lock in the upper hemisphere, they entered this time through a drab, dull orifice in the black half of the sphere. The patrol ship which brought them was contacted without ceremony. They were thrust though with curt orders to ask somebody for the Mug superintendent's office. Then the valve closed behind them. There was a grating sound as it was locked from the outside, and then silence. The ship was gone. They were marooned in the gloom, the grisly domain of the rays and the Mugs. Sentenced for life, with their only companions, a few broken, despairing men.

The corridor in which they found themselves sloped gently downward, and artificial gravity made it possible to walk naturally. Sine taking the lead, they passed into the depths. Everywhere were monstrous shadows, with occasional stabbing eerie beams of light. But it seemed that an ominous hush hung over this metal-interlaced gulf. Here there was no sense of motion—no sense of bubble-like lightness. It was like a descent into the nether regions of the ancient—into an inferno. But of the denizens of this dismal place there was no trace.

"Let's go to Proserpina's home," Sine suggested. "I'm anxious to see if she's still all right. And the old man too."

Accordingly they watched for the numbered corridor, and after some fruitless wandering, came again to the deep crack that was the only home this timid girl knew. She started up in terror as the Earthmen came into view. Not unnaturally, for they were all bristly with unshaven beards and grimy with the dust they had collected when prisoners in the Jovian ship's hold.

But after her first reaction of terror she gave a glad cry, and running up to Sine, threw her thin arms around his muscular neck.

"Now listen, kid!" The young scientist began with unwonted embarrassment. But the girl clung to him, and he could not quite bring himself to tear her arms away. She released him herself, in a few moments, became suddenly shy.

Lents laughed with genuine amusement.

"Don't be silly, Sine. She's just glad to see us again. Poor kid was lonesome. Come here, Prosie."

She went to him, gravely embraced him; then Kass.

They noticed she was trembling.

"What's the matter?" Kass asked. "You act as if you're glad to see us, but wished we hadn't come."

"Why are you here?" she asked with a troubled frown.

The Earthmen told her of what had transpired—that they were now condemned for life to serve in the dark hemisphere. As they spoke her fears seemed to vanish. She became radiant with delight.

"Then you have come at the right time!" she cried. "Our slavery is at an end, and you shall pilot us back to the Mother planet!"

"You're not crazy, are you kid?" Sine asked, lifting her little pointed chin with his hand.

"No!" she laughed delightfully. "Not crazy!" And she would have embraced Sine again. "My father has been building a ship for the past two years, hoping to escape to Ganymede, or some other moon of Jupiter. But now we shall go to Earth!" She clapped her hands excitedly.

"Listen! Let's get this straight," Lents demanded. "You say your dad has built a ship. Where is it?"

"Way down in the bottom of the hemisphere. That's where all the Mugs are, working on it when they have time. Dad's chest feels better again."

"They have built a ship, huh?" Sine was trying to suppress the hope that flamed up madly. "How'll they get it out?"

"They've made an airlock, so that when we leave the escaping air won't give us away."

It was one of these things that seem too good to be true. But when the Earthmen accompanied the girl to the secret workshop, directly next to the sphere's outer skin, they found she had spoken the truth in every respect. The men there, nearly all pathetic wrecks of the First Race's system, were at first a little doubtful about admitting the Earthmen, but one after another they were won over to the idea of seeking sanctuary on Earth rather than on some satellite of Jupiter where they would never be entirely safe. Besides, the Earthmen, though they had been stripped of all their weapons, represented additional fighting strength.

They made their final preparations with mixed feelings. Many of the Mugs had relatives on Jupiter, though few had wives or children. Even women of the Second Race had no desire to share the fate of a man condemned to a lifetime in the black half of the Bubble. Those few women who had accompanied their men to the metal satellite would, of course, be taken along, for the escape ship was commodious.

The next two weeks were filled with arduous labor, but at last the ship was ready, and observation through a small port which had been installed, showed that they were about to enter the shadow of Jupiter. Under cover of darkness they would leave the airlock. They would accelerate past The Bubble. Centrifugal force would send them away from Jupiter. At the same time their velocity with relation to the sun would be diminished. Lents plotted a long, graceful curve that would bring them to Earth with the best possible speed.

Proserpina's father lay on the floor, peering out through the port.

"Remember, Jan," Lents reminded him, "as soon as we cut the shadow, you give the order." They were all in the ship save the Earthmen and Jan, lying on the floor like a great spider, with his tremendous chest laboring painfully.

"In a moment now," Jan said. "The sun is nearing the limb."

"Open! Open, you aberrated spores!" The command came but faintly through the inside valve of the emergency airlock.

"They've found out!" Kass gasped. "Quick, never mind the shadow!"

Jan had already leaped to the long cylindrical hull, the product of endless labor and sacrifice.

"Inside!" Sine shouted. Kass and Sine made for the ship's manports. "I'll take care of the thermite."

In his hand he carried a small heat pistol that had long ago been stolen and hidden by a Mug. Quickly he made a circuit of the room, which was like an enormous sheet-metal blister on the inside of the metal satellite. After the thermite had cut out the ship free, that blister would prevent the escape of air, saving the lives of thousands of the First Race and also preventing discovery of their escape for a time.

The thermite was piled generously in a ridge all the way around. Sine leaped inside the first valve of the manport, colliding with a soft body.

"Get inside, kid!" He leveled his pistol at the thermite ridge where it was nearest to him. High time too. The walls of the blister were radiating heat. The fools were turning their infra-red beams on it!

"Lock!" Sine shouted, pressing the trigger and jumping back.

Instantly the ship was surrounded by an oval of brilliant orange and white fire. The valve clicked shut in Sine's face, and he dived through the second one into the interior, tripping the lock of that one also.

Through the ports nothing was to be seen now save fire. They were in an inferno of brilliant light and heat. But through the glare and smoke Sine saw a white-hot spot suddenly appear on the blister wall. The Jovians were melting their way through! The metal plates sagged like wet paper, dropped limply. Back of the hole, luridly illuminated, stood the foremost of a detachment of fighting men, eager to leap to the fray, waiting only for the metal to cool a little.

But the thermite had been burning steadily, biting through the tough skin of the metal moon. Just as the fugitives were beginning to wonder whether they would be incinerated in their self-made prison there was a lurch. Through the hull of their own vessel they could hear the tearing of metal as the weakened plates were sheared away. They found themselves in space, with the great ball of the Pleasure Bubble floating away from them. Just outside of the gaping hole in the sphere floated the bodies of twenty or thirty men, blown out by the escaping air.

The air was escaping in a prodigious geyser; unimpeded by an atmosphere, it spewed out, visible like a cloud due to its moisture, smooth like an inflating balloon without billows. The ball of vapor expanded swiftly toward the gray vastness of Jupiter 100,000 miles below, enveloping the fugitive ship for a time, then passing on, like an enormous milky white cloud, falling swiftly until it was lost in the darkness, still expanding.

Overhead The Bubble continued serenely on its course, the sweeping curve of its crystal hemisphere visible. But now the actinic lights that had served as artificial suns were dark. The great man-made paradise was as cold and dead as the Earth's moon. Death stalked its pleasure palaces. Already up there the pleasant rippling lakes must be skimmed over with ice, the luxuriant vegetation stiff, crackling with frost.

Despite the selfishness, the cruelty, the utter callousness of the First Race, Sine felt a pang of regret over the destruction of so much beauty.

A messenger from the astrogator's cabin, a man whose skin was seared and scorched so that it looked like an alligator's hide, touched Lents' arm.

"Jan would like to have you verify the course." There was apprehension in the man's voice. Member of a race so long enslaved, restrained, he feared the freedom of open space.

They swept slowly past The Bubble, gaining speed. Suddenly there was a cry from the stern look-out:

"The ship's heating. Stop it! Something's wrong."

Sine, rushing to answer the call, found that the ship was indeed heating up. Shielded from the sun's rays as they were, this was inexplicable. And then he saw the dull red pinpoint of light.

He had not seen it before, that patrol ship, clinging like a leech to the airlock of the crystal hemisphere. There had been men in there when the air escaped. They had been saved from death by the closing of their automatic airlocks.

"Better get back into the shelter of The Bubble," he told Jan after a hurried trip to the astrogator's cabin. The spider man turned the vessel, and they scurried back to shelter. Although the patrol ship tried its gravity buttons on them, the Mugs had fully equipped their own vessel with similar, and larger buttons which were occasionally used in regulating the metal satellite's orbit. They could neutralize the other vessel's gravity force with ease.

"And yet," Sine admitted to the serious little group in the cabin, as they once more floated in space under the immense sphere, "they seem to have us stymied."

"Suppose they follow us around here?" Kass asked somewhat nervously.

"I don't think they can," Sine said. "I noticed when we came to The Bubble first, the ships are locked to the gaskets from inside the sphere. The men inside the ship can not unlock their ship unless they open the emergency air curtain. If they did their air would all escape through the sphere. They could do it, of course, if they put on space suits. But that procedure would take an hour, and in the mean time we could get out of range of their heat rays. So we have them stymied too. Except for one thing——"

"Of course," Lents grunted. "We can't get at them, and they can't get at us, but in a few hours we'll be in sunlight again, and some patrol will pick us up."

The Mugs, watching fearfully from beyond the doorway, turned aside. Were they, after a mere glimpse of freedom, to be immediately returned to the bondage which had become unbearable to them? Sine felt a small, thin hand slip into his. He looked down into the wistful face of Proserpina looking up at him with hope, with confidence. All at once his shyness vanished as he realized that Proserpina's obvious adoration for him was only the admiration of a child for a very big and very wonderful brother. At the same time his desire to do something to release them all from their peril was intensified by the imperatively felt need to justify her confidence in him. An idea came.

"Jan," he asked. "What is the energy output—the total capacity—of our gravity buttons?"

Jan named an approximate figure in ergs.

"Lents, if you've ever calculated to a purpose, calculate now! How much energy is represented by the mass of that sphere at its orbital velocity?"

"I get you!" The fat scientist puffed out his cheeks with excitement. "Have to estimate the mass first." He picked up a stylus from the astrogator's table, worked furiously on a tablet. Kass and Jan watched apprehensively. The Pleasure Bubble, with its freight of the dead, was hurrying remorselessly to its rendezvous with the sunlight.

"Whoops!" Lents threw his tablet into the air in extravagant triumph. "She'll do!"

"Stations!" shouted Jan, in his curious strained voice, and men rushed eagerly to their posts, still hazy as to their object but cheered by the knowledge that there was hope after all.

Then began one of the strangest duels in the history of the solar system. Setting the nose of their vessel against the gigantic metal satellite, they directed the stern gravity buttons against a distant star, and applied full force to slow the sphere in its orbit.

The forces liberated were terrific. The sphere's tough skin, three inches thick, buckled and bent inward until the ship was almost buried in a pit of its own creation. Jan stood hunched over the activator lever like a great spider, ready to throw it into neutral at the first sign of an actual rupture, which would send them crashing through the internal cells and girders of the sphere.

"She's folding up like a squeezed orange peeling!" Kass muttered, running his hand over his bald head.

"Built to withstand internal pressure—nothing like this," Jan gasped. "Stout ship, this!" he added a moment later. "We thought we might have to ram our way out."

She was indeed a stout ship—this vessel of escape. Though she shivered and groaned, she gave no indication of failure.

"Wonder if the others are pushing against us!" Kass suddenly thought of another possibility.

"We—can—outpush 'em." Jan gasped. "Got to sit down. Here you take it!" Sine stepped into his place. Vague shocks and noises were transmitted to them through the hull. The huge sphere was collapsing progressively.

Lents came puffing from an observation port.

"She's slowing!" he reported triumphantly. "Our trajectory—give her a little more!"

The Joy Bubble was becoming more and more disc-shaped, and it was slowly turning on a major axis as the contending forces became uncentered.

"Flopping like a flapjack," Lents commented as he watched the shifting vista. A moment later; "It's a close squeeze. See there, past the horizon—a prominence?"

It was like a white plume, this jet of vapor thrown far into space. Not uncommon in Jupiter's turbulent atmosphere. But it was bright, dazzling! That meant they were not far from the sunlight!

"Pull away!" the fat mathematician shouted. "We have to take a chance!" Instantly Sine reversed the lever. Everyone grasped handholds as the ship backed out of the pit. Now they could see the vast ruin they had wrought. Sine gave her all the speed he dared, for the sun, for home!

The great ruin was slowly turning, and in a few minutes they saw again the darker shadow that was the fighting patrol ship, still clamped to her side. At the same instant the dull red pinpoint winked on. The Jovians had sighted them again! In a few minutes the hull was getting uncomfortably warm.

Lents laid down his pad.

"They will crash!" he declared. "But they have an hour, the fools! Instead of trying to burn us why don't they get into their space suits and free themselves?"

Jan, resting on the bench, shook his shaggy head.

"They are a great people, stupid but great. They will try to punish us till they die."

The wreckage drifted closer and closer to Jupiter, and still the red beam played steadily on the fleeing prisoners' ship. The distance had become so great that it could only be seen through an old telescope that the prisoners had somehow procured. But the prisoners were gasping. Their hull was cherry-red on the outside, and still heating. A few more minutes and the heat would be unbearable.

"They are getting closer—closer—they are in the sunlight. Now I can see better. I believe they will skip by—no! They've dived into the vapor! They're out again. Skipped out like a flat stone on water. Sinking again—almost over the horizon. Gone, I guess. Whew, it's hot!"

They were accelerating so fast that they had to turn on the interior gravity buttons to equalize the pressure on their bodies. Behind lay the vast, fog-bound planet of Jupiter. Ahead was the beautiful sun. And somewhere beyond, and still invisible, Earth the lovely, the green, the Mother of the human race!